MW00966601

At Sylvan, we believe that a lifelong love of learning begins at an early age, and we are glad you have chosen our resources to help your child experience the joy of mathematics to build critical reasoning skills. We know that the time you spend with your children reinforcing the lessons learned in school will contribute to their love of learning.

Success in math requires more than just memorizing basic facts and algorithms; it also requires children to make sense of size, shape, and numbers as they appear in the world. Children who can connect their understanding of math to the world around them will be ready for the challenges of mathematics as they advance to topics that are more complex.

At Sylvan we use a research-based, step-by-step process in teaching math that includes thought-provoking math problems and activities. As students increase their success as problem solvers, they become more confident. With increasing confidence, students build even more success. The design of the Sylvan activity book will help you to help your child build the skills and confidence that will contribute to success in school.

Included with your purchase of this activity book is a coupon for a discount at a participating Sylvan Learning center. We hope you will use this coupon to further your child's academic journey. To learn more about Sylvan and our innovative in-center programs, call 1-800-EDUCATE or visit www.SylvanLearning.com.

We look forward to partnering with you to support the development of a confident, well-prepared, independent learner.

The Sylvan Team

Published in the United States by Random House, Inc., New York, and in Canada by Random House of Canada Limited, Toronto.

www.tutoring.sylvanlearning.com

Producer & Editorial Direction: The Linguistic Edge
Writer: Amy Kraft
Cover and Interior Illustrations: Shawn Finley and Duendes del Sur
Layout and Art Direction: SunDried Penguin

First Edition

ISBN: 978-0-307-47946-4
ISSN: 2161-9689

This book is available at special discounts for bulk purchases for sales promotions or premiums. For more information, write to Special Markets/ Premium Sales, 1745 Broadway, MD 6-2, New York, New York 10019 or e-mail specialmarkets@randomhouse.com.

PRINTED IN THE UNITED STATES OF AMERICA

10 9 8 7 6 5 4 3 2 1

Sylvan
Learning sm

Kindergarten
Fun with Numbers

Connect the Dots

DRAW a line to connect the numbers in order, starting with 1.

Happy Clowns

DRAW balloons in each clown's hand to match the number the clown is holding.

Connect the Dots

DRAW a line to connect the numbers in order, starting with 1.

Happy Clowns

DRAW balloons in each clown's hand to match the number the clown is holding.

X Marks the Spot

DRAW an X on any group of bugs that has less than 5.

What's the Order?

WRITE the order of the pictures from 1st to 4th.

1

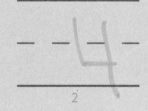

2

3

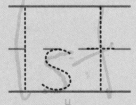

4

What's the Order?

WRITE the order of the pictures from 1st to 4th.

_____
_ 1 _
_____
1

_____
_ 3 _
_____
2

_____
_ 4 _
_____
3

_____
_ 2 _
_____
4

Snake Spiral

FOLLOW the pattern and COLOR the snake.

Incredible Illusions

FOLLOW the pattern to finish coloring the picture. LOOK at the lines both before and after you color. Do they look straight?

HINT: Use a marker instead of a crayon.

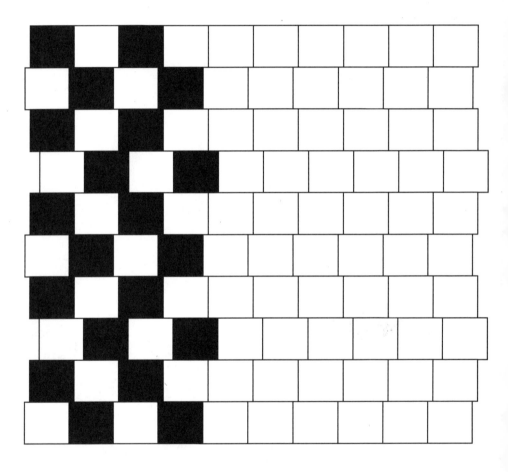

Get the Cheese

FOLLOW the path by connecting the numbers in order from 1 through 10 to get the cheese.

Dock the Boat

FOLLOW the path by connecting the numbers in backward order from 10 through 1 to dock the boat.

Picking Pairs

DRAW a line to connect each pair of objects that belong together.

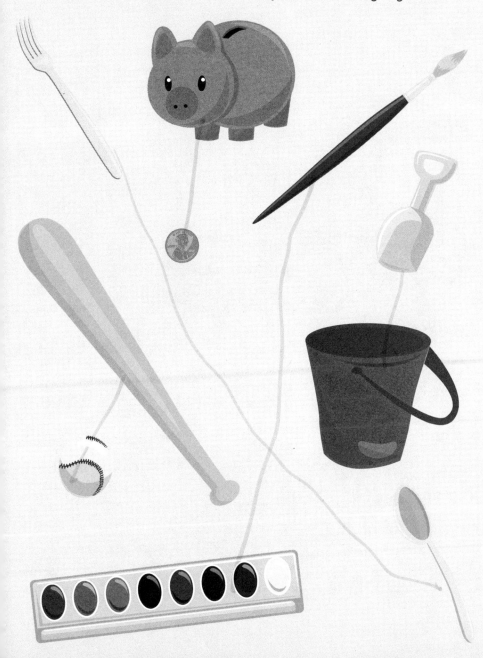

Zookeeper

The new zoo signs are in the wrong places. DRAW a line from each sign to the place where it belongs.

Picking Pairs

DRAW a line to connect each animal with its home.

Pick for Packing

Help Nolan pack for a camping trip. DRAW lines from the objects he will need to his backpack.

Cross Out

DRAW an X on any shape that is **not** a circle.

circle

Recognizing Shapes

Cross Out

Draw an X on any shape that is not a triangle.

triangle

Cross Out

Draw an X on any shape that is not a square.

square

Doodle Pad

TRACE the circles. Then DRAW a picture using each circle. Think about how many things are circles.

Doodle Pad

TRACE the triangles. Then DRAW a picture using each triangle. Think about how many things are triangles.

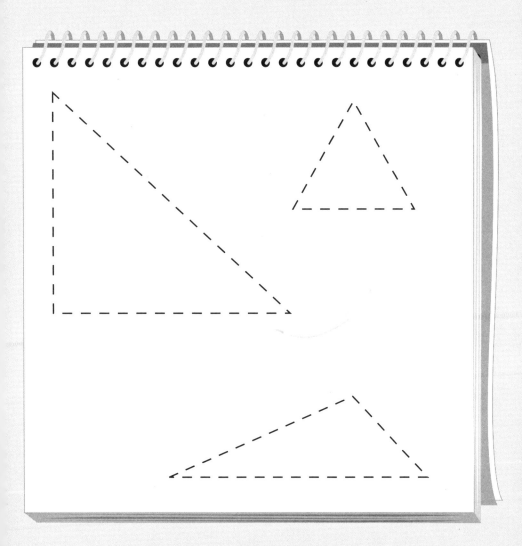

Doodle Pad

TRACE the squares. Then DRAW a picture using each square. Think about how many things are squares.

Doodle Pad

DRAW three things above the chair and two things below the chair.

Treasure Hunt

There are many pirate treasures, but only one is real. FOLLOW the pirate's directions, and DRAW an X on the correct treasure.

Walk from my straight for a . Go around

the past some . Then head for the

and cross a . There ye will find my .

County Fair

First prize at the Cooper County Fair goes to the longest carrot. CIRCLE the carrot that wins first prize.

County Fair

First prize at the Cooper County Fair goes to the heaviest pumpkin.
CIRCLE the pumpkin that wins first prize.

Pick for Packing

Help Ricky pack for his airplane trip. He can only bring things that will fit in his small suitcase. DRAW lines to his suitcase from each object that will fit inside.

Answers

Page 2

Page 3

Page 4

Page 5

Page 6

Page 7

1. 2nd   2. 4th
3. 3rd.   4. 1st

Page 8

1. 1st   2. 3rd
4. 4th   5. 2nd

Page 9

Page 10

Page 11

Page 12

Page 13

Page 14

Page 15

Page 16

Page 17

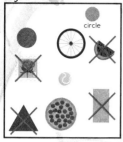

Page 18

Page 19

Page 20
Have someone check your
answer.

Page 21
Have someone check your
answer.

Page 22
Have someone check your
answer.

Page 23
Have someone check your answer.

Page 24

Page 25

Page 26

Page 27